# Fact #1

The tongue of a blue whale can weigh as much as an elephant!

# Fact #2

The fingerprints of a koala are so similar to humans that they can be mistaken for each other.

# Fact #3

Cows have best friends and can become stressed when they are separated from them.

# Fact #4

Elephants can communicate over long distances by producing low-frequency sounds that travel through the ground.

# Fact #5

The hummingbird is the only bird that can fly backward.

# Fact #6

A group of flamingos is called a "flamboyance."

# Fact #7

Cheetahs are the fastest land animals, capable of reaching speeds up to 75 miles per hour (120 kilometres per hour).

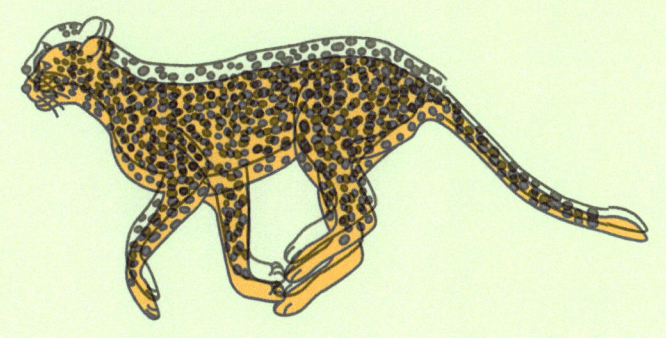

# Fact #8

The heart of a shrimp is located in its head!

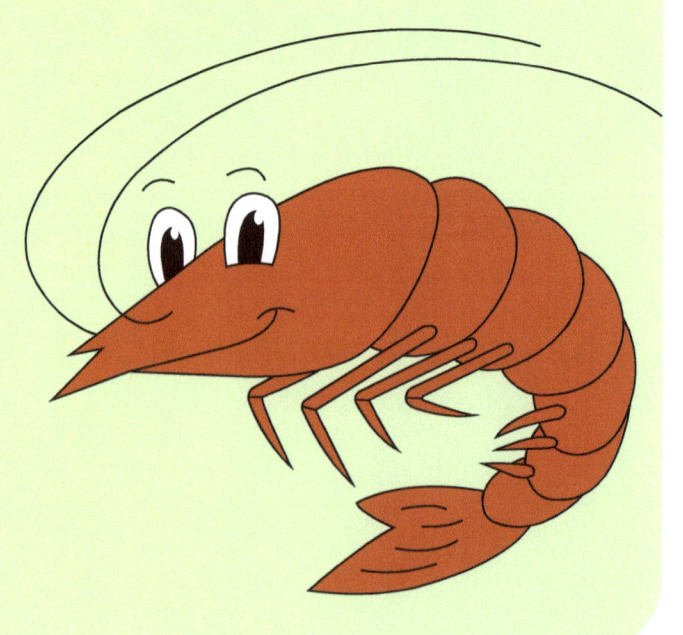

# Fact #9

Octopuses have three hearts and blue blood.

# Fact #10

Giraffes have the same number of neck vertebrae as humans: seven.

# Fact #11

Female lions do most of the hunting for their pride, while male lions typically guard the territory.

# Fact #12

Sloths can hold their breath for up to 40 minutes underwater.

# Fact #13

The tongue of a chameleon can be as long as its body!

# Fact #14

Honeybees can recognize human faces.

# Fact #15

The fingerprints of a koala are so similar to humans that they can be mistaken for each other.

# LET'S RECAP

What can weigh as much as an elephant on a blue whale?

Tongue

What features do koalas share with humans that can lead to mistaken identity?

# Fingerprints

What can cows become stressed about when separated from their companions?

# Their Friends

How do elephants communicate over long distances?

# Sounds

Which bird is capable of flying backward?

# Hummingbird

What is a group of flamingos called?

# Flamboyance

How fast can cheetahs run?

75 miles per hour

Where is the heart of a shrimp located?

# In the Head

How many hearts does an octopus have?

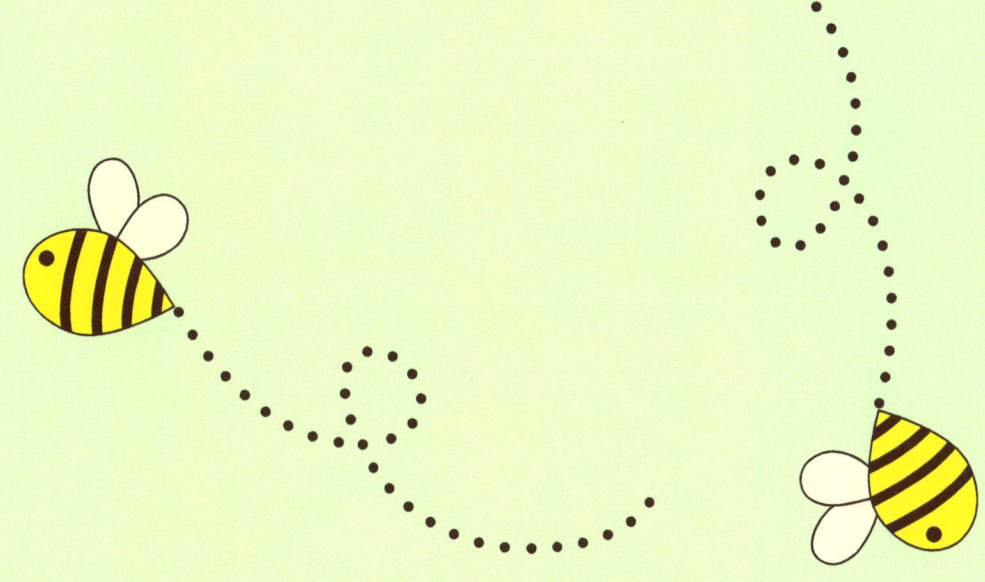

# Three

How many neck vertebrae do giraffes have?

# Seven

What is the primary role of female lions within their pride?

# Hunting

How long can sloths hold their breath underwater?

40 minutes

What can be as long as a chameleon's body?

# Their tongue

What unique ability do honeybees have in recognizing individuals?

# Faces

What feature of koalas can be mistaken for human fingerprints?

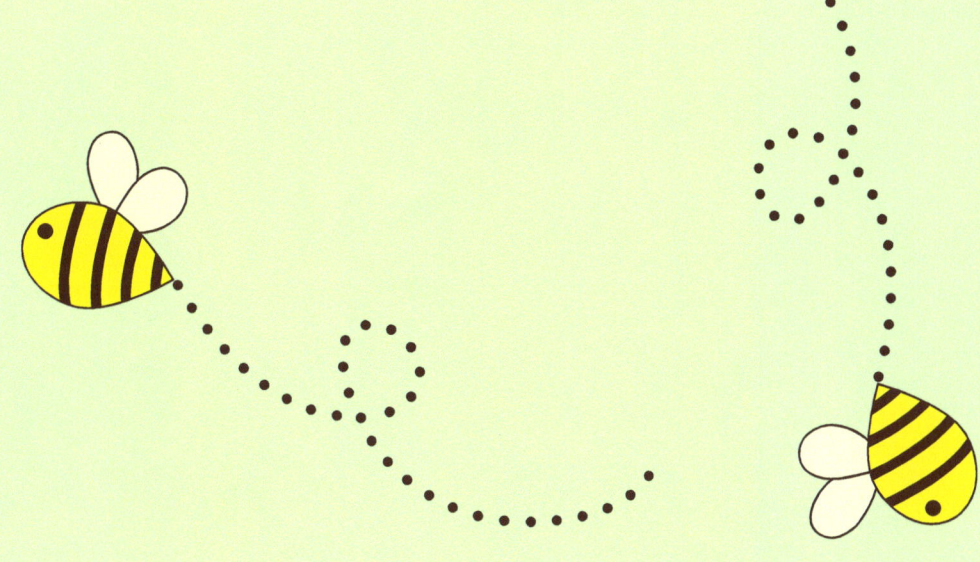

# Fingerprints

## ABOUT THE AUTHOR

Nisha is an educational professional with a fervor for storytelling and a background in science. She loves storytelling in her classrooms and loves to work on the social-emotional learning of young minds. She loves to create helpful content for learning and reading. She believes in making this world a better place by sensitizing people with better teaching-learning-knowing processes. Should you have any suggestions, write us at nishawrites18@gmail.com. Your feedback and suggestions are important to us. Happy reading.

www.ingramcontent.com/pod-product-compliance
Lightning Source LLC
Chambersburg PA
CBHW051928210526
45473CB00006B/2170